$2

D0894462

FOSS Science Resources

Living Systems

Full Option Science System
Developed at
The Lawrence Hall of Science,
University of California, Berkeley
Published and distributed by
Delta Education,
a member of the School Specialty Family

© 2016 by The Regents of the University of California. All rights reserved. No part of this book may be reproduced or transmitted in any form or by any means, electronic or mechanical, including photocopying or recording, or by any information storage and retrieval system, without prior written permission.

1487710
978-1-62571-354-4
Printing 3 — 11/2017
Quad/Graphics, Versailles, KY

Table of Contents

Introduction to Systems

Have you heard of the solar system? You probably know that it is the Sun, Earth, Moon, Mars, Venus, and all the other planets and their moons. But why is it called the solar *system*? Solar refers to the Sun, but what is a **system**?

A system is a collection of **interacting** parts. The parts work together to create a structure or produce an action. In the solar system, the parts include the Sun and the planets and their moons. The interactions are the motions of the objects and the force of gravity that holds them together. Gravity and the collection of objects work together as a system.

A pair of scissors is a system. It has two levers with sharp edges. Your hand applies force to the ends of the two levers. The cutting edges apply force to the material between the blades. This system cuts through the material.

The solar system is huge and complex. A pair of scissors is small and simple. Both systems are collections of interacting parts.

A bridge is a system.

A tower is part of a cell phone system.

A bridge is also a system of interacting parts. The parts include steel girders, steel cables, bolts, and concrete supports. The bridge system is a structure that is solid and stable. Bridges don't move much, but they support the weight of the traffic traveling across them.

If you look closely, you can find systems everywhere. Some systems are industrial, like a waste-water treatment plant. Some systems are social, like a hospital or school. Some systems are technological, like the cellular telephone system and the Internet. Some systems are natural, like the migration of monarch butterflies and the cycle of the tides. Some systems are cultural, like baseball. What systems can you think of?

Thinking about Systems

Here is a list of eight systems. Can you think of three or four parts of each system?

1. Refrigerator
2. Skateboard
3. Sunglasses
4. Helium balloon

5. Flashlight
6. Hot dog
7. Sleeping bag
8. Belt

Is Earth a System?

Our planet, Earth, is a really big object, with many interacting parts. Interacting parts! That sounds like Earth might be a system.

Earth has a core made of iron and nickel. Around the core is a thick layer of rock called the mantle. The outside surface of Earth is covered by a thin, hard, rocky crust. The core, mantle, and crust are a system called the **geosphere**. But there is more to Earth than just the geosphere.

Much of Earth is covered by water. Most of the water is in the ocean. A lot of water is in rivers, streams, lakes, ponds, and underground. And more water is stored in huge masses of ice in Greenland, Antarctica, and glaciers. Water is in the air as invisible water vapor and as clouds and fog. The interacting water on, under, and above Earth's surface is a system called the **hydrosphere**.

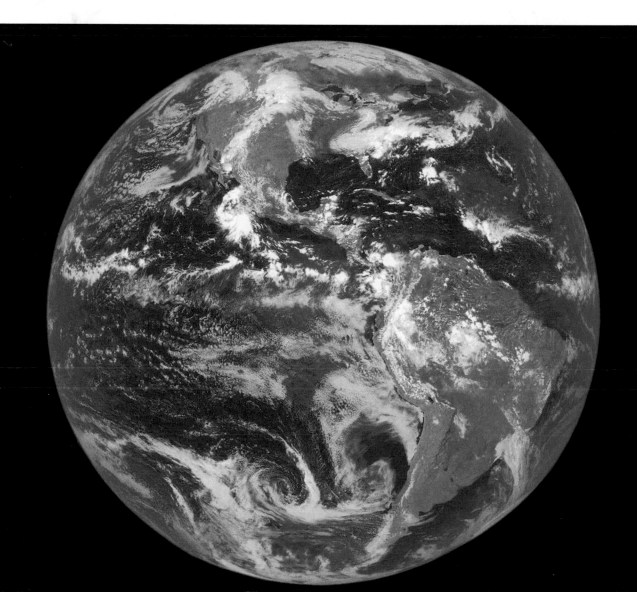

Above Earth's surface is the **atmosphere**. The atmosphere is a system of interacting gases called air.

Earth's crust, ocean, lakes, ponds, streams, and air are home to millions of plants and animals. This is the **biosphere**. The biosphere is a system of interacting living organisms.

So yes, Earth is a system. The huge Earth system includes the geosphere, hydrosphere, atmosphere, and biosphere. Perhaps the most interesting part of the Earth system is the biosphere. Think about it for a second. What are the parts of the biosphere? What are some of your interacting parts? How do they work together to make the biosphere?

Living organisms in a tide pool are part of the biosphere.

Earth's atmosphere

The Biosphere

The biosphere is all of the organisms living on Earth. Some of the interacting parts live in the North Atlantic Ocean **ecosystem**. The North Atlantic organisms are very different from the organisms living in a coral reef ecosystem or in a desert ecosystem. Everywhere you go, you find different kinds of ecosystems. Each ecosystem has its own kind of organisms.

An ecosystem is part of the biosphere. Every ecosystem has thousands of interacting parts. Each kind of organism is a part of the ecosystem in which it lives. For instance, the Sonoran Desert ecosystem includes saguaro cacti, mesquite trees, Gila woodpeckers, elf owls, horned lizards, sphinx moths, milkweed bugs, harvester ants, kangaroo rats, coyotes, and many other organisms. Organisms interact with one another in many ways.

Food Chains and Food Webs

Food chains and **food webs** are systems of interacting organisms. A food chain simply describes a feeding relationship among a few organisms. A food web is a more complex system showing all the feeding relationships in an ecosystem.

Organisms that make their own food are called **producers**. In **terrestrial** ecosystems, the most important producers are plants. Grasses, trees, and bushes, are producers. In freshwater and ocean ecosystems, algae and **phytoplankton** are the most important producers. Producers make their own food from sunlight, water, **minerals**, and **carbon dioxide** (CO_2).

Animals are **consumers**. Consumers get their food by eating plants or other animals. Plant or animal material that is not eaten is consumed by **bacteria** and **fungi**. Bacteria and fungi are **decomposers**. After decomposition, only minerals are left. The minerals help produce the next generation of plants.

Food Chains

When a spider eats a fly, the matter and **energy** in the fly go to the spider. This feeding relationship can be shown with an arrow. The arrow always points in the direction that the matter and energy flow.

fly   **spider**

If a praying mantis eats a spider, the matter and energy in the spider go to the praying mantis.

fly   **spider**   **praying mantis**

It's possible in a woodland ecosystem for a blue jay to eat the praying mantis, a weasel to eat the blue jay, and a hawk to eat the weasel. Matter and energy pass from one organism to the next when they are eaten. This is called a food chain. And at the beginning of the food chain is a producer. Energy for producers comes from the Sun.

In this case, the producer is a fruit from a tree, a plum. You can draw arrows from one organism to the next to describe a food chain. The arrows show the direction of energy flow. They point from the organism that is eaten to the organism that eats it.

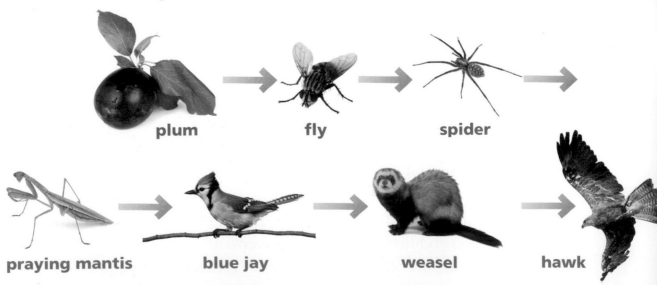

plum   **fly**   **spider**

praying mantis   **blue jay**   **weasel**   **hawk**

Another example of a food chain might have grass as the producer. A chipmunk eats the grass seed. A hawk eats the chipmunk. Bacteria decompose any dead organisms or uneaten parts. You can always draw arrows from dead organisms to the decomposers.

A simple food chain

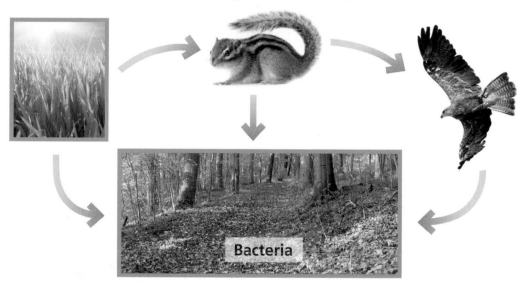

Bacteria

Food Webs

There are many feeding relationships in an ecosystem. If you draw *all* the arrows that show who eats whom, you have a food web, not a food chain. The food web for a freshwater river might look like this.

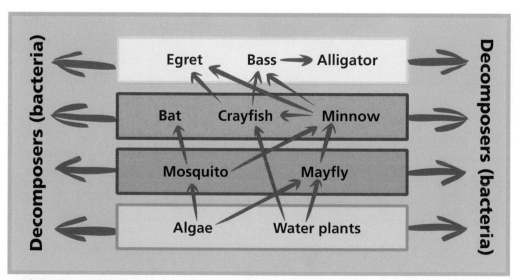

This is an example of a food web for a freshwater river. Bacteria decompose all the organisms when they die.

Locate the crayfish in the example of a food web. Crayfish are food for both egrets and bass. If the river has a lot of crayfish, egrets and bass will both have plenty to eat. But if there are few crayfish, the egrets and bass will have to **compete** with each other for food.

The animal that can get more food is the one that is more likely to survive. In this river ecosystem, egrets and bass compete for crayfish. Are there other competitions for food in the ecosystem?

Organisms in ecosystems depend on one another for the food they need to survive. Herbivores depend on producers to make food. Carnivores depend on consumers for food. Omnivores depend on both producers and consumers for food. Decomposers depend on dead organisms and waste for food. And producers depend on decomposers for raw materials to make food. In a healthy ecosystem, some organisms will be eaten so that other organisms will survive.

Egrets eat crayfish.

Your body has many different systems for different functions.

Your eye is part of the system that helps you see.

Human Systems

You are an organism, living in an ecosystem, interacting with many other organisms in many ways. That means you are part of a system. But are *you* a system?

You do many different functions, including moving, seeing, hearing, eating, smelling, tasting, thinking, talking, and breathing. Each function is performed by a different system in your body.

When you run across the schoolyard, two systems are interacting. Your skeleton is a system of 206 strong, hard bones in many sizes and shapes. Attached to your bones is a system of hundreds of muscles. The bones and muscles work together as you run.

Seeing uses a system that includes your eyes, which convert light into electrical pulses. The electrical pulses travel along nerves into your **brain**. Hearing is a similar system, including an outer ear, inner ear, nerves, and the brain. Smell and taste have systems of **receptors**, nerves, and brain centers.

As you can see, you are a system of subsystems. And you are a subsystem in an ecosystem, which is a subsystem of the Earth system. And Earth is one planet in the solar system, which is one planetary system in the Milky Way Galaxy. The universe is an endless system of subsystems.

Thinking about the Biosphere

1. What is a simple definition of the biosphere on Earth?

2. What is an ecosystem?

3. What is one way that organisms interact in an ecosystem?

4. What is the role of producers in an ecosystem?

5. How might human actions affect the food web in a woodland or a freshwater river?

Monterey Bay

Monterey Bay National Marine Sanctuary

Much of the northern California coast is rocks and cliffs. The ocean water is very cold all year. During the winter and spring, huge waves from the Pacific Ocean crash on the rugged shore. Can anything live in this difficult environment?

The answer is yes. The northern California coast is one of the most diverse and productive ecosystems on Earth. Thousands of different kinds of organisms live and interact in the cold ocean water. This ecosystem is protected in the Monterey Bay National Marine Sanctuary. *Marine* means "ocean" or "sea." A sanctuary is a protected place. This is one place where scientists can study the interactions between ocean organisms and their environment.

The Kelp Forest

Giant kelp grows in most of the 15,783-square-kilometer (km) sanctuary. Kelp looks like a plant, but it is actually algae. Like plants, algae make their own food.

Giant kelp are anchored to the seabed and reach clear to the ocean surface. In some places, the distance is more than 100 meters (m) to the surface. This makes the kelp taller than the tallest trees. For this reason, the California marine ecosystem is often called the kelp forest.

Like the rain forest, the kelp forest has a floor, an understory, and a large canopy. The canopy spreads across the water's surface. But, unlike the rain forest, most of the organisms do not live in the canopy. Most live in the understory and on the floor. Every bit of the rocky bottom has animals clinging to it. These include clams, scallops, mussels, barnacles, limpets, abalones, snails, sponges, sea urchins, sea stars, shrimp, and sea anemones. Every crack and cave shelters a fish, an eel, a crab, or an octopus.

A kelp forest

Fish live in the understory. There are small fish such as anchovies and sardines, medium-sized fish such as sea bass, snappers, and perch, and large fish such as groupers and sharks. The California state marine fish is the bright orange garibaldi. It also lives here. Other animals found in the understory are squids, jellyfish, seals, sea lions, and gray whales.

The canopy provides shelter for a number of small animals that live on and around the kelp. These include snails, crabs, barnacles, and kelp fish. The canopy is a resting and hunting place for sea otters, seabirds, gulls, terns, ospreys, and ducks.

Where do all these animals get the food they need to survive? Like all ecosystems, the kelp forest depends on producers. The giant algae provide matter and energy to the ecosystem, but only a small amount. Microscopic phytoplankton are the most important producers in this ecosystem. These tiny producers (the grass of the sea) are eaten by **zooplankton**. Zooplankton are eaten by baby fish (kelp fish), clams, crabs, and thousands of other organisms. Small fish and crabs are eaten by larger fish (sea bass). The food produced by the phytoplankton eventually feeds the sea lions and sharks at the top of the food web. Marine bacteria decompose all the dead organisms in the ocean ecosystem.

An orange garibaldi

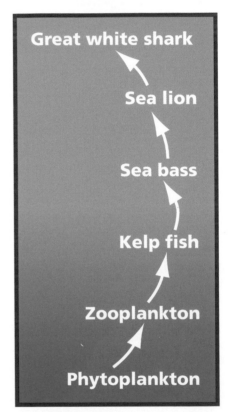

Great white shark

Sea lion

Sea bass

Kelp fish

Zooplankton

Phytoplankton

Monterey Bay food chain

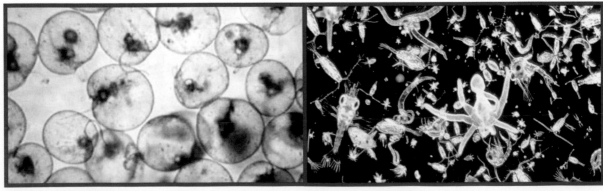

Phytoplankton　　　　**Zooplankton**

Competition for Resources

There is a lot of competition for phytoplankton in the marine ecosystem. The zooplankton that have the best structures for catching phytoplankton will be most successful. This is one example of competition for food.

There is also competition for space. Waves and currents are very strong in the coastal environment. Many organisms must attach firmly to a solid surface or be washed away. The rocky bottom of the ocean is completely covered with organisms.

Kelp forest organisms compete for shelter. Caves, cracks, and old shells are used as hiding places. There is life-or-death competition for places to attach and hide.

This is a hermit crab. Hermit crabs live in empty snail shells. What kind of competition do you think they have in the ecosystem?

Thinking about Marine Ecosystems

1. What do you think happens to waste and dead animals in marine ecosystems?

2. What is the most important producer in both freshwater and marine ecosystems?

3. Identify three ways organisms compete in marine ecosystems.

A terrestrial ecosystem

Comparing Aquatic and Terrestrial Ecosystems

Aquatic and terrestrial ecosystems are very different. But they are the same in some ways. Let's compare.

The **nonliving** factors of the two environments are different. **Aquatic** ecosystems are in water. Terrestrial ecosystems are on land. The temperature in an aquatic ecosystem changes slowly. The temperature in a terrestrial ecosystem can change rapidly over a short period of time. The amount of water in an aquatic ecosystem is predictable. Water in a terrestrial ecosystem can vary widely.

The organisms are different in the two ecosystems. Most aquatic organisms can live only in water. If they were moved to a terrestrial ecosystem, they would die. The same is true for terrestrial organisms moved into aquatic ecosystems.

An aquatic ecosystem

A heron is a consumer of crayfish in an aquatic ecosystem.

A fox is a consumer of mice in a terrestrial ecosystem.

Both ecosystems, however, are organized in similar ways. The organisms in aquatic and terrestrial ecosystems all need matter and energy to stay alive.

- Both ecosystems obtain energy from the Sun and matter from the environment.
- Both have food chains and food webs.
- Both have consumers that depend on producers to make food.
- Both have decomposers that break down dead organisms and recycle the raw materials (**nutrients**).
- Herbivores, carnivores, omnivores, and scavengers live in both ecosystems.

In both ecosystems, organisms compete for the resources they need to survive. Plants compete for light. Animals compete for food. Organisms need space and shelter from predators and changes in the nonliving environment. The organism that outcompetes the others is the organism that will survive.

Nature's Recycling System

Think of a tree. Like any organism, the tree will eventually die and fall to the forest floor. What happens to it? Does it pile up with other dead trees, plants, and animals, year after year?

When a tree falls in the forest, it is used for food by decomposers. Organisms that feed on dead trees are called **detritivores**. Some detritivores, such as beetle larvae and worms, dig into the trunks and eat the dead bark and wood. As they eat through the wood, the tree starts to fall apart. Other detritivores, such as termites, dig in and consume more of the wood. As the wood is exposed, fungi and bacteria move in. They consume the last of the wood and the waste left behind by the first decomposers. After several years, all that remains is minerals.

Animal bones, dead leaves, twigs, and fruit are organic matter called detritus.

Let's look more closely at the recycling system. In the deciduous forest of the eastern United States, **detritus** is most visible in fall. Then you are sure to see a layer of dead leaves and twigs, a few large tree limbs, and whole fallen trees. You might see a feather or a clump of fur left behind by a bird or raccoon, or scat (a pile of animal waste). You might find seeds and fallen fruit. You could find a piece of snake skin, or the bones of an animal. All of these bits of organic matter are detritus, and detritus is part of every healthy ecosystem.

You might think that detritus is waste and trash. But decomposers use this accumulation as food. The first decomposers to use the detritus are the detritivores. They concentrate on the largest parts of the detritus layer. Animals like termites, beetle larvae, isopods, and worms start to eat the fallen leaves, and dead wood. As they eat the dead matter, the mass of detritus decreases slowly. The detritivores leave waste of their own, which becomes detritus.

Termites eat wood and other detritus.

Detritus worms are different from earthworms. Detritus worms live in the dead leaf layer. Earthworms burrow into the soil, where they live under the detritus layer. One common detritus worm is called the redworm, or red wiggler. Home composters use redworms to decompose kitchen waste into rich fertilizer for gardens.

After the detritivores have chopped everything up, the real decomposers (bacteria and fungi) get to work. Bacteria and fungi work at the microscopic level. Bacteria don't have **mouths**. Instead, they leak chemicals onto the detritus. The chemicals dissolve nutritious materials from the detritus. The bacteria soak up the dissolved materials. Then they move on to the next bit of organic matter to repeat the process. Fungi get their nutrients the same way. When the fungi and bacteria are finished, the detritus has been reduced to simple minerals. The bacteria soak up the dissolved nutrients. Plants use these minerals to produce food. Then, the system starts again.

Redworms are detritus worms used in composting.

Fungi decompose dead trees.

Mushrooms are fungi.

There's Yeast in My Bread!

If you ever watched someone make bread, you might have noticed that they added a light brown material called yeast to the dough. What is yeast, and why add it to the dough?

Yeast

Yeast is a kind of fungus. You might have heard of some other kinds of fungus. Mildew is a kind of fungus that grows on organic materials like paper and leather. Mushrooms are the visible part of fungi that live on organic matter in the soil. These fungi are distant relatives of the yeast used to make bread.

One baker's yeast organism is a single **cell**. A single yeast cell is way too small to be seen with unaided eyes. With a microscope, you can see that one yeast organism is a tiny round object.

Why put fungus in your bread dough? Yeast eats **sugar**. When a yeast cell takes in a molecule of sugar, it breaks it down to use it for energy. The yeast breaks several carbon atoms off the sugar molecule. The carbon atoms combine with **oxygen**, forming carbon dioxide (CO_2) gas. The carbon dioxide produced by the yeast creates thousands of tiny bubbles in the dough. The dough rises as it fills with gas bubbles. The bubbles make the bread light and soft.

Where does the sugar that feeds the yeast come from? Some bread recipes call for a little sugar, but extra sugar is not necessary. Wheat flour contains a lot of starch and a small amount of a chemical called an enzyme. The enzyme breaks down the starch molecules into simpler molecules. Some of these molecules are sugar. The sugar that the yeast eats is from the starch in the flour. It takes time for the enzyme to act on the starch. So it can take several hours for bread to rise.

Next time you have a slice of bread, look closely at its texture. It is all full of holes, like a sponge. The holes were carbon dioxide bubbles. And remember, when you are eating a piece of bread, you are eating millions of baked yeast cells. Yum!

The holes in bread are the result of carbon dioxide bubbles.

Producers

Plants produce their own food. The food is sugar. The sugar is used by all plant cells. The cells use the energy in the sugar for growth.

Plants use a process called **photosynthesis** to make sugar for growth. The raw materials that plants use to make the sugar are water and carbon dioxide (CO_2). Water from the soil and carbon dioxide from the air combine with light energy from the Sun. Sugar, oxygen, and water are the products.

Most plants are green. Or at least they have a lot of green leaves. Leaves look green because the leaf cells have **chlorophyll**. Chlorophyll can absorb blue and red light. It reflects green light. That's why chlorophyll looks green.

The important part is that chlorophyll absorbs blue and red light. The energy from the absorbed blue and red light is then used to make the sugar molecules during photosynthesis. Sugar is the energy nutrient used by the plant cells to perform their life functions.

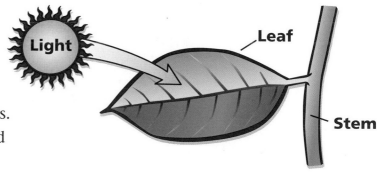

The green leaf cells make sugar out of carbon dioxide (CO_2) and water (H_2O). Carbon dioxide comes from the air. Water comes from the soil, up through the roots. The carbon dioxide and water meet in the green cells.

The carbon dioxide, water, and energy from the Sun combine to make sugar molecules in the plant's cells. The cells also produce oxygen and water molecules. The oxygen is released into the air. The plant reuses water or releases it into the air as water vapor (gas). So where is food produced? Food is produced in the green parts of the plant. Every cell that contains chlorophyll is making sugar.

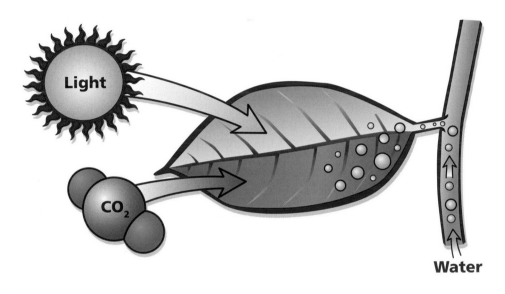

Water from the roots, carbon dioxide from the air, and light from the Sun enter the cells.

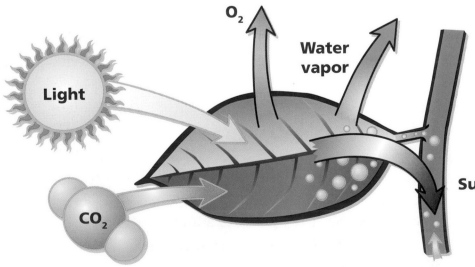

Carbon dioxide, water, and light combine to make sugar. Oxygen (O_2) and water are released into the air.

The Carbon Dioxide-Oxygen Cycle

Through the process of photosynthesis, producers release oxygen to the air. This is very important to the biosphere. Animals and most other living things need oxygen to live. Even plants need oxygen for their life functions. Producers release more oxygen than they use. Carbon dioxide is released as a waste gas by animals and other living things. Producers use that gas to produce more oxygen. This carbon dioxide-oxygen cycle is very important to the health of the biosphere.

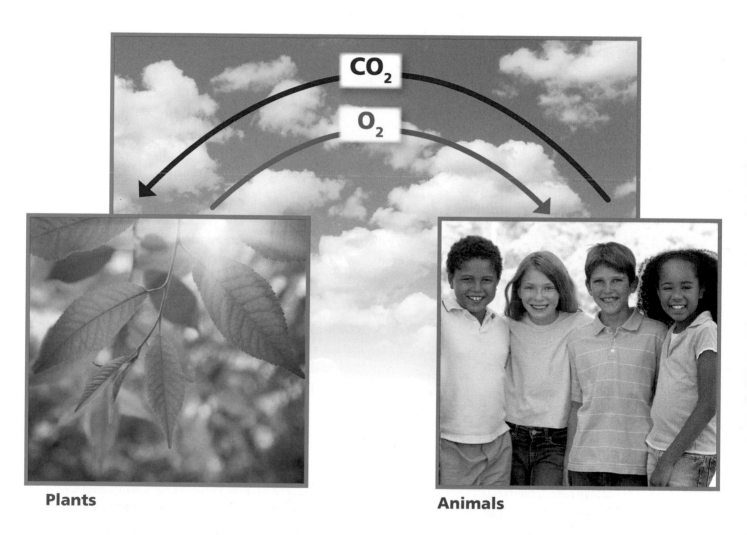

Plants

Animals

Food production doesn't stop there. Plants use the sugar to produce a lot of other molecules. They produce other kinds of sugars with names like sucrose, dextrose, and fructose. They produce starches, which store energy in potatoes and grains like wheat. They produce vegetable oils, such as corn oil, sunflower oil, and olive oil.

Plants store energy as sugars, starches, and oils. When the plant needs them, it pulls them out of storage, turns them back into **glucose**, and sends the glucose to the cells. That's how plants survive at night and during winter. They use stored energy to do whatever they need to do.

Other organisms use the energy stored by plants to live and survive. That includes humans. When you eat a slice of bread or a baked potato, you are eating energy stored by a plant. When you eat lettuce and carrots, you are eating sugars, starches, and all the cells made by plants. And when you eat food to nourish your cells, remember where the food came from. It started as carbon dioxide, water, and sunlight. It's really quite amazing when you stop to think about it. You are running on solar energy.

The Sun's energy is used by producers to make their own food. Then, that energy is transferred to you when you eat a plant.

Thinking about Photosynthesis

1. What is sugar?

2. What raw materials do plants need for growth? Where do those materials come from?

3. What is the role played by chlorophyll?

4. What are the products of photosynthesis? Where do they go?

5. Where do plants produce food?

6. Explain how the Sun's energy is transferred through a simple food chain.

Getting Nutrients

All animals, fungi, and many bacteria consume other organisms to get the nutrition they need to live and survive. These organisms are called heterotrophs. Plants, algae, and some bacteria produce their own food, so they do not need to eat other organisms. These producers are called autotrophs. Heterotrophs get their nutrients by eating other organisms or parts of organisms, alive or dead, for food.

Food is important for two reasons. It provides building blocks for growth, development, and system repair. And food is the source of energy that organisms need to live.

Autotrophs produce their own food.

Heterotrophs consume plants and other animals.

An adult butterfly

Butterfly Nutrition

Butterflies start life as a tiny egg. When the egg hatches, the tiny larva, called a caterpillar, must eat. Every kind of butterfly has a particular kind of plant that it uses for food. Painted lady larvae feed on mallow plants. The mallow plant is an autotroph. It produces food from carbon dioxide (CO_2), water, and sunlight. The leaves are made of **carbohydrates**, lipids, and proteins, nutrients that the caterpillar needs to live. The caterpillar nibbles off bits of leaf with its biting jaws and swallows them. The caterpillar's gut digests the leaf bits. **Digestion** releases the nutrients that the caterpillar uses to grow. The caterpillar grows and grows, laying in a supply of fat.

When the caterpillar reaches full size, it finds a proper location, attaches itself, hangs down, and pupates. Inside its protective covering, the caterpillar changes into its flying phase. The insect does not eat during this change. It uses energy and matter stored in its body to construct wings, legs, and a new system for feeding.

Butterfly life cycle

After a couple of weeks, the hard outer **membrane** splits. The adult butterfly climbs out and flexes its wings. After pumping fluid into the wing veins, the new painted lady can fly. The adult needs to feed in order to survive. The painted lady's **digestive system** has changed completely. The painted lady no longer has biting jaws for nibbling on leaves. Its mouth has changed into a long, thin tube called a proboscis. The tube is used to suck sweet nectar from flowers. Nectar is a good source of sugar. Sugar provides energy for the butterfly. Flying requires a lot of energy, so access to an energy-rich food source improves the butterfly's chances of survival.

While the butterfly is going about its business, all of the other organisms in the ecosystem are going about their business, too. Animals in the ecosystem are looking for food. The blue jay is always alert for his next meal. If he spots a painted lady larva munching on a mallow leaf, he will likely swoop down and gobble it up.

Blue jays eat butterflies.

Human Nutrition

How do *you* get your food? You are a player at all levels of a food pyramid. When you eat spinach, carrots, apples, or green beans, you are eating producers. Animals that eat producers are primary consumers, like humans and cattle. When you eat a piece of roast beef, you are eating a primary consumer.

When you eat a sardine, you are eating a secondary consumer. Secondary consumers eat primary consumers. Sardines eat little primary consumers called zooplankton such as copepods and fish and crab larvae. Zooplankton eat producers called phytoplankton. If you have a piece of salmon, you are eating a third-level consumer. The salmon eats the sardine (a secondary consumer). So, when you eat the salmon, you are acting as a fourth-level consumer.

Salmon

Sardines

Zooplankton

Phytoplankton

Humans eat at many consumer levels.

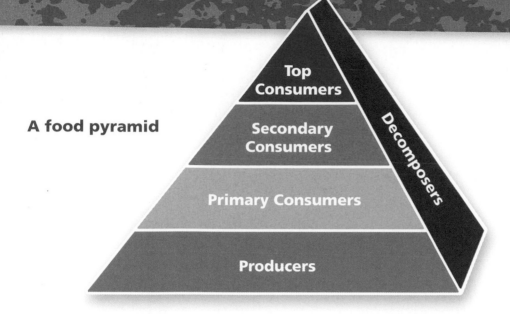

A food pyramid

Humans are aggressive top consumers, like tigers, sharks, orcas, and eagles. But unlike those animals, we can also eat lower on the food pyramid.

How do you extract the nutrients you need from your food? You eat to feed the trillions of living cells that make up your body. The process of breaking human food into nutrients for cells is called digestion. Cells get energy and raw materials from three groups of nutrients. They are carbohydrates, fats, and proteins.

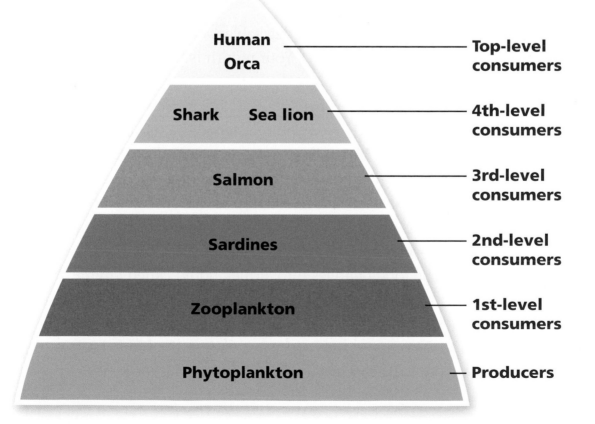

The Human Digestive System

Turning cheese, crackers, meat, vegetables, and fruit juice into nutrients for cells starts in your mouth. Your mouth is the beginning of a disassembly line for food. **Teeth** cut, mash, and grind up large pieces of food. **Saliva** mixes with the food to get it wet and to help break down the food. When you have chewed and moistened the food, you swallow it.

A wad of food, called a **bolus**, leaves the mouth and starts down the **esophagus** toward the **stomach**. Muscles along the length of the esophagus contract to push the bolus along. Your stomach is not just a place where a meal is stored. Things get rough down there. Digestive juices, including acid, are added to the food. Muscles in the stomach squeeze and mash the food. The food changes into a thick liquid called chyme.

The chyme moves into the **small intestine**, which can be 6 meters (m) long. More digestive juices are added. The small intestine has many bacteria. They attack and decompose the food you ate. Here the food changes into nutrients that your cells can use.

The small intestine is lined with millions of **capillaries**. Nutrients pass through the walls of the intestine into the capillaries. The blood carries the nutrients throughout your body, providing building blocks and energy for cells.

The last bits of the food move from the small intestine into the **large intestine** and **colon**. By this time, most nutrients are gone. Bacteria in the colon break down the remaining usable food. Water is extracted also. The remaining material contains fiber, other indigestible material, and dead bacteria. It is called feces. The feces moves into the rectum and is eliminated through the anus.

Because humans are animals, we cannot make our own food. We have to eat food to get our nutrients. Every cell in **multicellular organisms** needs nutrients. The digestive system breaks complex food sources into simple chemicals (nutrients). Those simple chemicals enter the blood and are transported to all the cells.

Food provides the nutrients our bodies need to survive.

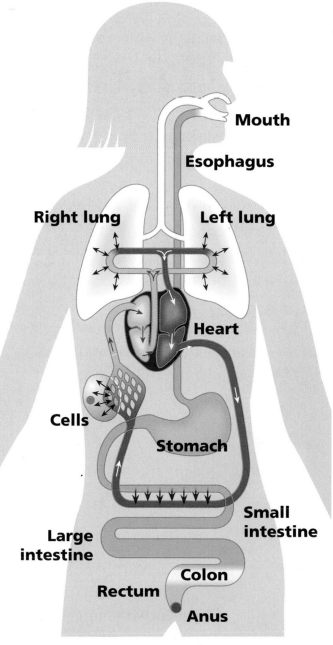

Mouth

Esophagus

Right lung

Left lung

Heart

Cells

Stomach

Large intestine

Small intestine

Colon

Rectum

Anus

Leaf Classification

A Japanese maple tree

Most **vascular plants** have leaves. The leaves on one kind of plant are different from the leaves on other kinds of plants. Scientists can use leaves to identify plants. But with so many different kinds of plants in the world, how do scientists use leaves to identify plants? The answer is **classification** systems.

Leaves have properties that can be used to organize them into groups, or classes. In class, you used the pattern of veins in the leaves to organize your leaf collection. You organized the leaves into three classes, **palmate**, **pinnate**, and **parallel**.

Palmate

Pinnate

Parallel

Other properties can be used to **classify** leaves, too. Leaves have shape. Some are long and pointed. Others are round. Leaves can even be fan shaped, triangular, or heart shaped. You can classify leaves by the shape of the **blade**.

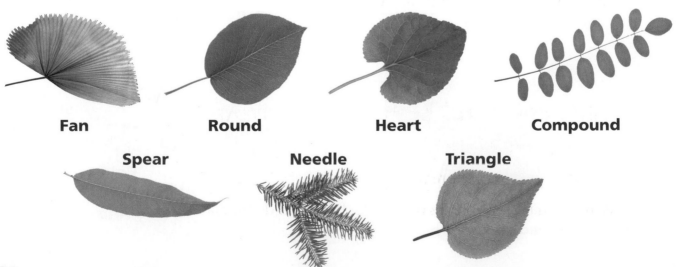

Fan　　**Round**　　**Heart**　　**Compound**

Spear　　**Needle**　　**Triangle**

The edges of leaves, called **margins**, are different from one another. Margins can be toothed, lobed, scalloped, fuzzy, or smooth. You can classify leaves by their margins.

Toothed   **Lobed**   **Scalloped**

Fuzzy   **Smooth**

Leaves are not the only way to classify plants. Whole plants can be classified. They can be organized into grasses, clovers, cacti, sagebrushes, palms, and so on. Any collection can be classified. Rocks can be classified by the minerals they contain or by form. A collection of rocks can be divided into a set that contains mica, a set that contains calcite, a set that contains quartz, and so on. The same collection of rocks can be classified again into sets of igneous, sedimentary, and metamorphic rocks.

Classification is one way to organize information about the natural world. By putting things together that have the same properties or behaviors, the complex world becomes a little easier to understand.

Thinking about Leaf Classification

1. What is classification?

2. What are three different ways you can classify leaves?

3. If you had a collection of insects, how would you classify them?

Plant Vascular Systems

General Sherman is the biggest living organism in the world. General Sherman is the name of a giant redwood tree living in Sequoia National Park in California. This giant tree stands over 85 meters (m) tall and is 11 m wide at the base.

Like all living organisms, General Sherman is a system made of living cells. Every cell needs water, nutrients, gases, and waste removal. How do all of General Sherman's billions of cells get the resources they need to survive?

General Sherman is a vascular plant. *Vascular* means "containing vessels." You have vessels called **arteries** and **veins**. Many plants have vessels, too. Other vascular plants include wildflowers, sagebrush, cacti, orange trees, lettuce, strawberries, wheat, peas, and celery. All vascular plants have a system of tubes running through them. These **specialized structures transport** nutrients to all the cells.

General Sherman is the world's largest living organism.

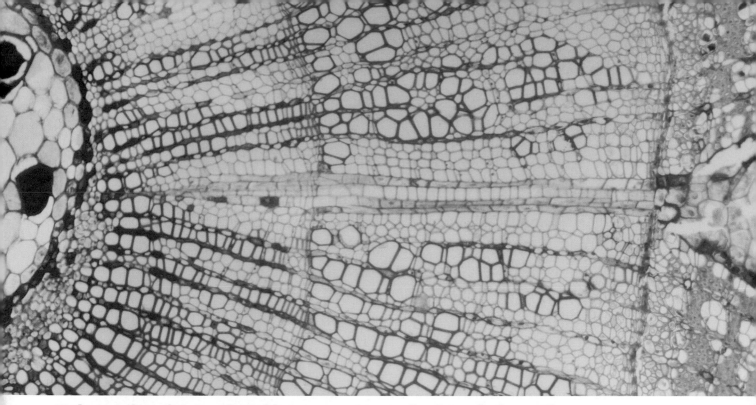

Xylem tubes (stained pink) carry water and minerals from the roots to the cells in the plant.

Xylem

Vascular plants have roots that reach deep into the soil. The roots take up water from the soil. The water enters long, hollow tubes called **xylem**. The xylem tubes start as long cells that are connected end to end. When the tubes are complete, the cells die. The resulting tubes transport water and minerals to the cells at the very top of General Sherman and to all the other living cells as well.

If you cut across the trunk of a tree, you can see the xylem tubes. New xylem cells grow all the time. The old xylem tubes form the main trunk of the tree. We call the old xylem cells wood.

Old xylem cells can be seen as rings in tree trunks.

37

Phloem

The green leaves of plants produce sugar. The sugar is the food used by all the cells in the plant. Some cells, like root cells and flower cells, do not make sugar. They need to get sugar from the cells that make it.

Vascular plants have a second kind of tube called **phloem**. Phloem tubes transport a sugar-rich liquid called **sap**. The phloem delivers sugar to every living cell that cannot make its own sugar.

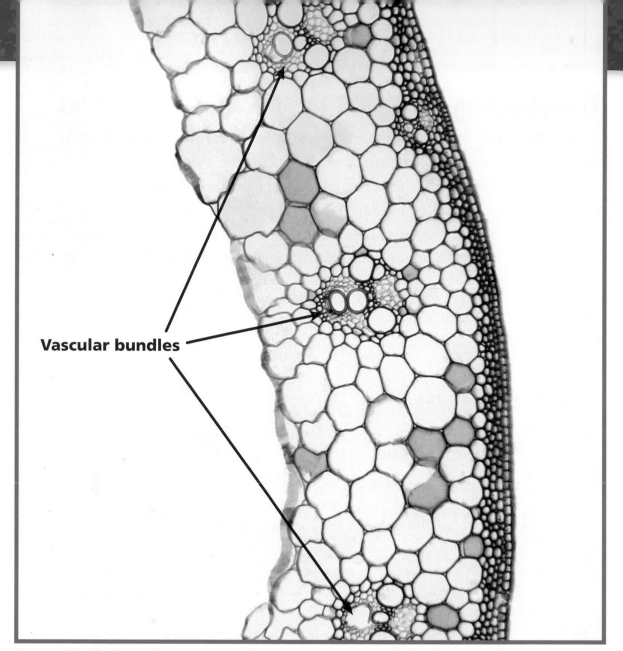

Vascular bundles

A portion of a wheat stem cross section showing vascular bundles (100X)

Many vascular plants have specialized structures called **vascular bundles**. A vascular bundle includes xylem tubes and phloem tubes. A celery stalk has vascular bundles you can see in cross section. With a microscope, you can see the xylem and phloem bundles in other plants, like wheat.

The dark areas at the edge of this celery cross section are vascular bundles.

Transporting Nutrients to and from the Leaves

Vascular plants have two systems of transport tubes that work together. Tubes carrying water up from the roots make up the xylem system. Tubes carrying nutrients down the plant make up the phloem system. The system of xylem and phloem in vascular plants is something like the system of arteries and veins in humans. Take a close look at the illustration below. You can see how the xylem and phloem transport water, minerals, and nutrients to and from cells.

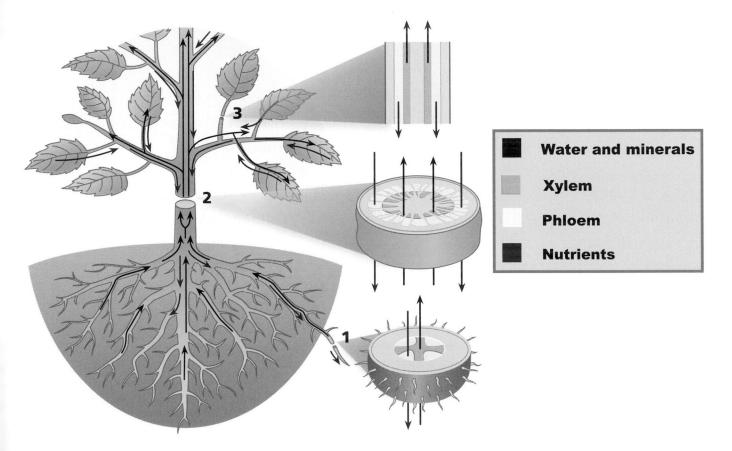

Water enters the roots underground. At number 1, a short section of root was cut. The section is shown enlarged. The pink tissue is the xylem. Water and minerals dissolved in the water flow up the root toward the stem. The black arrow shows the direction that water and minerals move through the xylem tubes.

At number 2, a section of the main stem was cut. The section has been enlarged. The xylem from all the roots passes through the stem. Often a group of xylem cells is near a group of phloem cells. They form a vascular bundle. You can see a lot of vascular bundles like spokes on a wheel around the outside of the stem.

At number 3, a section of leaf stem was cut. The section was cut again from top to bottom. In the enlarged view, you can see the xylem carrying nutrients and water to the veins in the leaves and from there to the cells in the leaves.

The xylem tubes end in spaces between the cells in the leaves. Minerals and some of the water are taken in by the cells. The rest of the water evaporates through tiny holes in the leaves and passes into the environment. This process is called **transpiration**.

Transporting Sugar to the Cells

Some green plant cells make more sugar than they need for energy. Extra sugar passes out of the cells into the tiny phloem tubes. The sugar mixes with water to make a sweet liquid called sap. The sap flows through the phloem to all the cells that are not green. Cells that are not green can't make their own sugar.

Look at the vascular-plant illustration again. This time follow the red arrows. From the leaf, the sugar flows through the tiny leaf stem (number 3) into the branches. The phloem in all the branches comes together in the main stem (number 2). Finally, the phloem branches out into all the roots, delivering sugar to all the cells in even the tiniest root (number 1). Every cell receives sugar so it can stay alive and do its job.

Xylem and phloem transport water and nutrients to the cells of plants.

Comparing Plants and Animals

Multicellular animals and vascular plants have specialized vascular systems to transport nutrients. In both plants and animals, nutrients flow through systems of vessels. But the systems in animals are different from the systems in plants. Animals have one system of vessels. Blood flows from the **heart** to the cells in arteries. Nutrients transfer to the cells in the capillaries. Then the blood returns to the heart in veins. Blood goes around and around, transporting everything cells need.

Plants have two systems of vessels that are not connected. Water flows from the roots through xylem tubes to all the cells. The water carries minerals as it goes. Extra water then evaporates into the air. Water passes *through* the plant. It does not **circulate** like the blood in animals.

Water and sugar (sap) come out of the green cells and flow to all the other cells in phloem tubes. The phloem carries food and water for cells.

Plants have two "one-way" systems. One system transports water and minerals up, and the other system transports nutrients down. Animals have one system that goes around and around.

Plants like this saguaro cactus have two systems of vessels.

Animals like this frog have one system of vessels.

The Story of Maple Syrup

Do you like to eat pancakes or French toast? These treats sure do taste yummy when covered in maple syrup. Real maple syrup comes from maple trees. Do you live in a place that has maple trees? There are more than 30 species of maple trees distributed across the United States. Are they all a source of maple syrup? Let's find out.

Making Maple Syrup

The first step in making maple syrup is to tap a maple tree to collect lots and lots of sap. Sap is basically water with some sugars and a few other substances in it. The sugars are produced during photosynthesis.

In fall, sap can flow into hollow tubes in the vascular system in and near the bark of the tree. There, the sugar in the sap is converted into starch for storage over the winter. As spring approaches, nighttime temperatures still drop below freezing, but daytime temperatures rise above freezing. Then the sap starts to flow. The starch changes into sugar and moves into the sap, which starts to flow down the tree trunk. If you drill a small hole into a tree and insert a small tube, the sap can drip out and into a bucket. The drip stops at night and starts up again as the day warms. The sap will continue to flow until the nighttime temperatures stay above freezing.

How do you turn this slightly sweet sap into very sweet syrup for pancakes? How do you get rid of the extra water? You boil it? Boiling evaporates the water, leaving the sugar behind. In fact, if you boil it too long, you will have a pot filled with solid sugar crystals. A lot of energy is needed to reduce the sap to syrup. Sap with higher sugar content requires less boiling than sap with lower sugar content. Using less energy is better for the environment and for farmers. It takes about 150 liters (L) of sap to produce 4 L of syrup.

The sugar maple tree, which grows in the northeastern United States, has a sugar content of 2–2.5 percent. Maple syrup is mostly made from the sap of sugar maple trees because it is the sweetest. Black maple, red maple, and silver maple also produce sweet sap worthy of boiling down. Other maple trees produce sap but have much lower sugar content, making it more difficult to produce syrup.

Sap has a lower freezing temperature than water: the higher the sugar content, the lower the freezing temperature. So the higher sugar content of the sap in trees in the cold parts of the country helps keep the sap from freezing. Trees in the warm parts of the country do not need sap with really high sugar content.

Sugar maple trees produce syrup.

The sap is boiled down to remove extra water to create syrup.

The steps for collecting sap

Trees and Sap

What is going on inside these giant living systems? You have already learned that trees have a vascular system with xylem and phloem. The tree's vascular system is in the sapwood, which is the part of the bark closest to the wood. The vascular system in a tree transports nutrients, carbohydrates, minerals, and water throughout the organism. Xylem transports water and minerals from the roots of the tree to the leaves. Phloem transports nutrients (sugars), produced by photosynthesis in the leaves, down the tree toward the roots. Trees use the sugar for energy to conduct the many activities that keep the tree alive and healthy. Trees produce more sap than they need for their own use, so people can tap some of it to use for maple syrup without injuring the trees.

What technologies do sugar farmers use to get the sap out of the trees? Maple sugar farmers tap their trees by drilling a hole about 5–6.5 centimeters (cm) into the trunk at a slightly upward angle. A spile (rhymes with smile) is tapped into the hole and a bucket is hung from the spile. A spile is kind of like a metal straw. On large farms tree tappers use a network of plastic tubing, often many kilometers long, to bring the sap from hundreds of spiles directly to the sugar shack. This saves a tremendous amount of time and effort collecting sap buckets.

How did the first person discover that maple trees have sweet sap? How did they get it out of the trees before drills were invented? Historians are in agreement that the Native Americans were the first to discover this sweet liquid.

To make sugar, Native Americans made a cut in the maple tree bark and collected the sap as it dripped out. They used hollow logs to store sap. They boiled the sap by dropping very hot stones into it. They eventually shared this technique with the European settlers. Over time, new technologies were invented. The settlers learned to bore holes in the maple trunks and to insert a wooden or metal spile. They used wooden buckets to catch the sap.

Birch and walnut trees can also be tapped for sap. Some Russians drink birch sap. In Ukraine, Japan, China, and South Korea, people drink raw sap for its health benefits.

Can you tap maple trees in your neighborhood and make maple syrup? Remember that to produce sap, the tree has to go through a very cold winter with many months of freezing weather. If your climate doesn't have harsh winters and slow transitions to spring, your trees won't produce lots of high sugar sap. The northeastern part of the United States and the southeastern part of Canada are the regions where most maple syrup is produced.

Many years ago it took 75–115 L of sap to make 4 L of syrup. Now it takes 151–189 L. That's a lot more sap. It also means a lot more boiling. Why is this happening? Scientists are still trying to answer this question. Maybe you'll become a scientist who studies trees. Let us know if you figure out what's going on with the sugar maple tree sap!

Tree sap is a natural resource. Can you find out if people tap other kinds of trees to make things other than syrup?

This family is tapping a maple tree for syrup.

Thinking about Maple Syrup

1. What technologies are used to collect and produce maple syrup?

2. What vascular tissue is tapped to collect the maple sap?

3. How has maple syrup changed in recent years?

4. How is pancake syrup different from maple syrup? Collect labels and promotional information from several brands of both syrups, and evaluate the information.

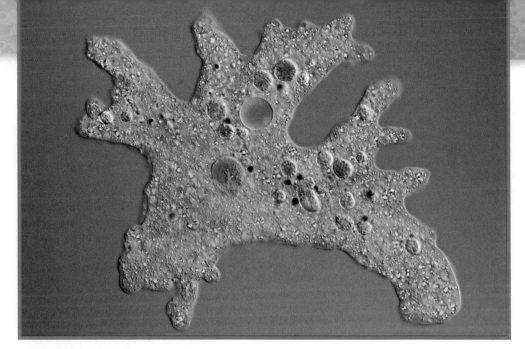

A living amoeba

The Human Circulatory System

The basic unit of life is the cell. All organisms are made of living cells. The simplest organisms, such as the amoeba, are just one cell.

All living cells have something in common. They all have a membrane on the outside. All cells are filled with a liquid called **cytoplasm**, which is mostly water. And all cells need four resources in order to stay alive. They are water, food, gases, and waste disposal.

How Do Cells Get Resources?

Single-celled organisms live in water. The food and gases they need to survive are in the water. The environment brings water, food, and gases to the cells all the time. The cell releases waste products into the water. The environment provides all the resources that single-celled organisms need.

A human is a multicellular organism. A human is made of trillions of cells. Humans don't live in water, and most of the cells are deep inside the body.

Muscles are made of millions of cells. Every cell in a human muscle is alive. That means every cell is getting the resources it needs to survive. How do these muscle cells get the water, food, gases, and waste removal they need to survive?

Multicellular organisms have specialized structures to transport resources to cells. In humans, blood, which is mostly water, is pumped through blood vessels to all the cells. The blood carries food and gases to the cells, and carries away wastes.

The human body is made of many different kinds of cells. There are nerve cells, muscle cells, bone cells, liver cells, lung cells, skin cells, and so on. A group of cells of the same kind, working together to perform a function, is called a tissue. Muscle tissue contracts to produce movement. Bone tissue gives our bodies structure. Nerve tissue sends electric messages. Each tissue is made of its own kinds of cells. But the cells in all tissues need the same basic resources.

Cells break down sugar to get energy. Cells need oxygen to do the job. One of the by-products of the sugar breakdown is the waste gas carbon dioxide (CO_2). If cells don't get oxygen, they will die. If cells don't get rid of the carbon dioxide, they will die.

The human body is made of many different kinds of cells.

Resource Delivery

Blood flows through blood vessels to every cell in the body. The blood is kept flowing with a pump called the heart. The human heart is a four-chambered organ made of powerful muscles. The muscles contract to pump the blood about once every second. You can feel the beat of your pumping heart when you put your hand on your chest. The heart muscle works all the time. It pumps day and night, year after year. Every year your heart beats more than 30 million times!

Blood flows away from the heart in blood vessels called arteries. Blood flows back to the heart in vessels called veins. The smallest blood vessels, the ones that serve the cells, are called capillaries. The system of blood vessels and the heart is called the **circulatory system**. It circulates blood to every cell in your body.

The human circulatory system

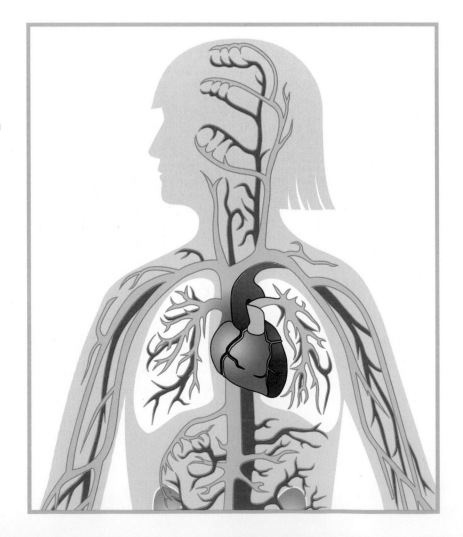

The two most important resources transported to cells are oxygen (a gas) and sugar (food). The most important waste product removed from cells is carbon dioxide (a gas). Oxygen comes from the air we breathe, and sugar comes from the food we eat. In order to get fresh oxygen, dispose of carbon dioxide, and get new sugar for cells, the circulatory system has to connect with the **lungs** and intestines.

To learn how the circulatory system works, let's take an imaginary trip through it. Red blood cells carry oxygen to the cells and carbon dioxide away from the cells. You have about 25 trillion red blood cells in your body. They live only about 4 months, so they are being replaced at the amazing rate of about 3 million per second.

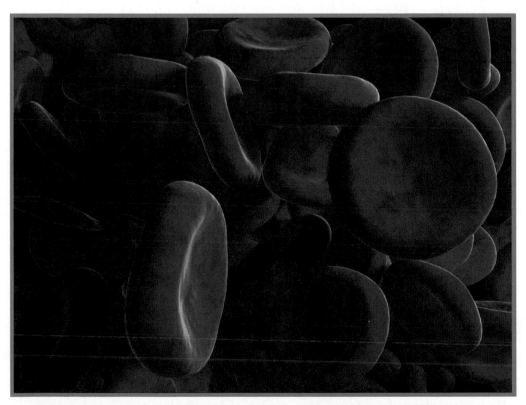

Red blood cells (scanning electron microscope view, about 4,000X)

The Right Side of the Heart

It takes about a minute for a red blood cell to travel once through the circulatory system. Blood returning from the body cells goes to the right side of the heart. The returning red blood cells are carrying carbon dioxide waste. The returning blood enters the upper chamber on the right side of the heart, called the **right atrium**. When the heart beats, the right atrium squeezes blood down into the **right ventricle**.

The next time the heart beats, it pushes blood out of the right ventricle to the lungs. The blood flows through tiny capillaries that are touching the air sacs in the lungs. The red blood cells release carbon dioxide. The carbon dioxide enters the air in the lungs and is exhaled. Then the red blood cells take oxygen from the air you breathe in.

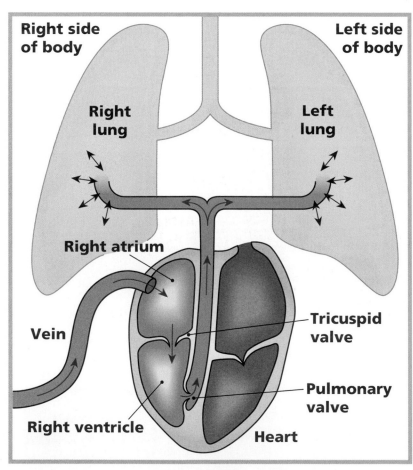

A diagram showing how the right side of the heart works

The Left Side of the Heart

The oxygen-rich red blood cells go back to the left side of the heart. Blood from the lungs flows into the **left atrium**. The next time the heart beats, it squeezes blood into the powerful **left ventricle**. When the left ventricle contracts, it pumps blood through arteries to the body. The red blood cells transport oxygen and pick up waste carbon dioxide. Then the cycle starts over again.

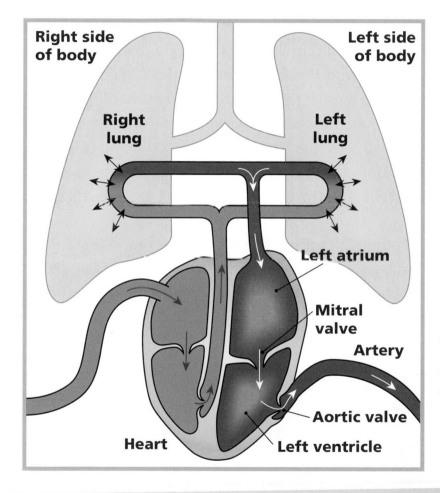

A diagram showing how the left side of the heart works

Thinking about the Human Circulatory System

1. What is the heart and what is its role in the circulatory system?

2. What are heart valves and what do they do?

3. Where are the heart valves?

4. What is the main function of the left side of the human heart?

5. What is the main function of the right side of the human heart?

The Human Respiratory System

The **respiratory system** has three main parts. They are the lungs, the system of tubes that connect the lungs with the outside air, and the diaphragm (an arched muscle). The respiratory system brings oxygen to the red blood cells and gets rid of waste carbon dioxide.

When your arched diaphragm muscle contracts, you breathe in. When you inhale (breathe in), oxygen from the air enters your lungs. The air ends up in the 300,000,000 alveoli (air sacs) at the ends of the tiny tubes (bronchioles) in your lungs. The alveoli are surrounded by capillaries. The oxygen passes through the walls of the air sacs into the capillaries. Red blood cells pick up the oxygen. At the same time, the red blood cells release waste carbon dioxide from the body cells into the alveoli. This waste gas goes into the air when you exhale.

Blood flows to the body tissues through arteries. The blood flows through smaller and smaller arteries, ending in networks of capillaries. Capillaries are only 1/100 of a millimeter in diameter. That's just a little bit larger than a red blood cell. Capillaries are so small that red blood cells often travel single file to get through.

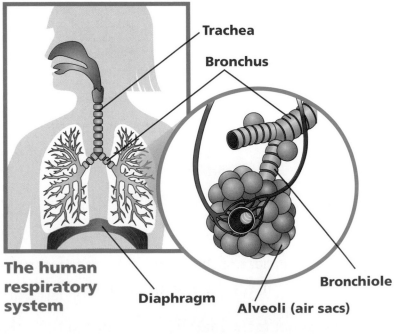

The human respiratory system

Trachea

Bronchus

Diaphragm

Alveoli (air sacs)

Bronchiole

The capillaries touch every cell in the body. Gas exchange takes place while the red blood cell is sliding past a cell. Here, only the thin wall of the capillary is between them. Oxygen passes into the cells, and carbon dioxide passes out. The red blood cell then transports the carbon dioxide to the lungs for disposal.

Red blood cells carry gases. They carry the essential gas, oxygen, to the cells and carry the waste gas, carbon dioxide, away from the cells.

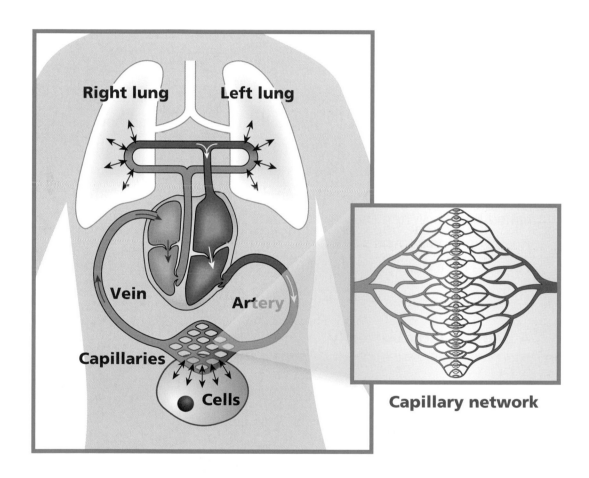

Capillary network

Thinking about the Human Respiratory System

1. What are the parts of the respiratory system? What is the system's function?

2. What are alveoli and what happens there?

Other Circulatory and Respiratory Systems

You have a closed circulatory system. A system of vessels and organs holds your blood, and it stays in there. One of the organs is a muscular heart, which pumps the blood around and around through the system. All vertebrate animals have a similar circulatory system. These animals include mammals, birds, reptiles, fish, and amphibians.

Invertebrate animals have different kinds of systems for distributing food and other nutrients to their cells. The painted lady butterfly is a member of the class of organisms called insects. Insects have an open circulatory system. One vessel runs from their head to the end of their abdomen. In the abdomen, the vessel forms several chambers called hearts. When the insect moves, its muscles make the hearts push the blood toward the head. The blood spills out into the body cavity. There it seeps freely around the organs and other tissues.

A painted lady butterfly

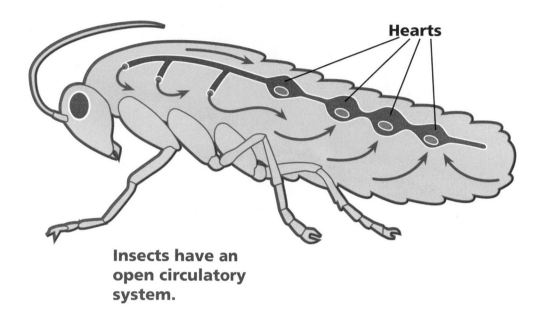

Hearts

Insects have an open circulatory system.

When the blood flows around the gut, nutrients from digested food enter the blood. The nutrient-rich blood nourishes all the cells as it flows through the insect's body cavity. The blood slowly flows toward the abdomen, and when it arrives at the area around the hearts, it enters the vessel through little valves in the hearts. Then the circuit repeats.

Insects don't have blood like you. Insect "blood" is called hemolymph and is yellowish or greenish. Hemolymph does not carry oxygen to the cells or carry waste carbon dioxide (CO_2) away from the cells. Insects have a completely different respiratory system. Air enters the insect's body through a line of holes along its side. The holes are connected to air tubes that branch and branch, eventually reaching all the cells in the insect's body. The cells get their oxygen from these little air tubes, and eliminate their waste carbon dioxide there, too.

Thinking about Other Circulatory and Respiratory Systems

Compare the structures and functions of the human circulatory system with that of the painted lady butterfly.

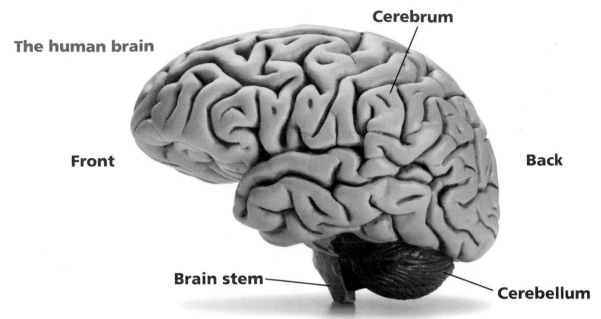

The human brain

Cerebrum

Front

Back

Brain stem

Cerebellum

Stimulus and Response in Humans

Structures of the Brain

The human brain is a compact mass weighing about 1.5 kilograms (kg). The spinal cord extends from the brain down through a hole in the backbone. The brain and spinal cord make up the **central nervous system**. The central nervous system is completely enclosed in bone. The brain is surrounded entirely by the cranium, or skull.

The brain has three major parts, the cerebrum, the cerebellum, and the brain stem. The largest part of the human brain is the large, bumpy, folded cerebrum. It has two halves, the right and left hemispheres. The spheres are symmetrical and are connected to each other. The cerebrum makes up about 70 percent of the mass of the whole human nervous system. The lobes of the cerebrum have areas that have specific functions.

It also has a distinctive set of folds. The fold patterns are similar for all humans. This folded outer surface is called the cerebral cortex. The cerebral cortex processes the signals that come into the brain. Without this thin surface layer, we would not be able to think, recognize faces, or plan ahead.

The more folds a brain has, the more it can process. The cerebrum of a rat, for instance, is smooth, implying that it is not a big thinker. The brain of a dolphin, on the other hand, is more folded than a human's.

The small, roundish structure that lies below and to the back of the brain is the cerebellum. It makes up about 11 percent of the mass of the brain. It processes information from the muscles, tendons, and inner ear. It uses this information to manage and maintain balance and coordination.

At the center of the brain is a small, cordlike structure called the brain stem. It connects the brain and the spinal cord and relays messages to and from the cerebrum and the cerebellum. The brain stem regulates many body functions, such as heartbeat, breathing, and body temperature. You can survive damage to the cerebrum or cerebellum, but damage to the brain stem is usually fatal.

The brain stem also relays information from the body to other parts of the brain. In general, the right hemisphere of the brain receives from and sends messages to the left side of the body, and vice versa. The brain stem coordinates the crossover.

A diagram comparing a human brain with a bird brain

Brain Messages

If an ant is walking on your arm, you know it even if you don't see it. Its feet tickle you, and without even looking, you raise your other arm to brush the ant away. How are you able to do that?

Your arm has touch receptors for the sensation we know as tickle. The ant walking on your arm is a **stimulus**. When a tickle receptor is stimulated, it sends a message to your brain, alerting you to a problem. Your brain decides how to **respond**. It sends a message to your arms, telling them what to do to brush away the ant.

The special cells that make up your brain and the rest of your nervous system are **neurons**. You have several hundreds of billions of neurons throughout your body and brain. Those neurons are constantly sending messages from one place to another.

Your touch receptors, photoreceptors (light sensors), and hearing, taste, and smell receptors are all on the ends of neurons. These **sensory neurons** send messages from the environment to the brain. The brain decides what to do about these messages. If your brain decides that you should act, it sends out messages to your muscles or other systems, telling them to snap into action. This call to action is sent on **motor neurons**.

The receptors on the ends of neurons tell your brain when something tickles.

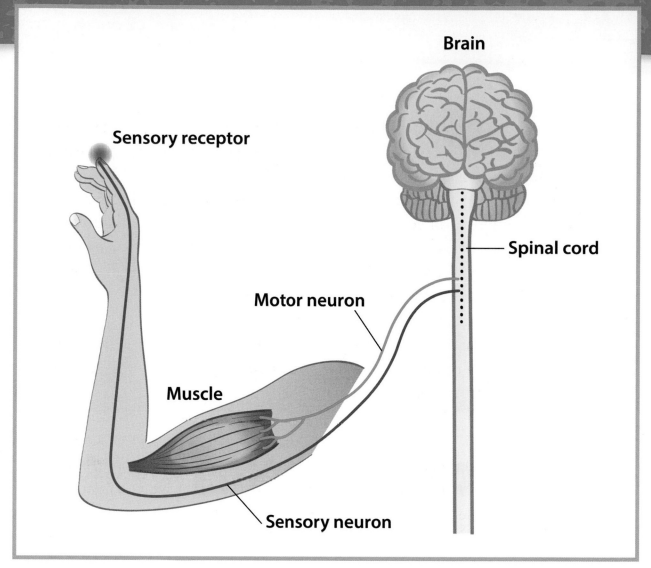

Brain

Sensory receptor

Spinal cord

Motor neuron

Muscle

Sensory neuron

Motor and sensory neurons keep your body in constant communication with your brain.

Sensory neurons and motor neurons are like wires carrying an electric signal. Sensory neurons carry messages to the brain, and motor neurons carry messages away from the brain. Sensory neurons give the brain information, and motor neurons send instructions to the muscles. Your arm responds to the message from the motor neurons by contracting certain muscles. Sensory neurons called stretch receptors give the brain feedback and tell it how much the muscles are stretched or contracted. This communication between the brain and muscles is happening constantly, all over your body.

Sending messages takes time. The longer the pathway, the longer it takes to produce a response. The interval is called response time. You might have noticed this delay when you stub your toe; you can see it being stubbed and hear the sound before you feel the pain! The pathway from your eyes and ears to your brain is much shorter than the pathway from your toes to your brain. So the sensory neurons in your eyes and ears get their messages to the brain before the sensory neurons in your toes can.

Sensory Systems

Awareness of the environment and the ability to respond quickly are absolutely essential for you to stay alive. Fortunately, you have been supplied with an early-warning system to tell you about potential hazards. The system is your senses and your brain, which controls your every action.

Staying Alive

Senses pick up clues from the environment, both far and near, and pass them on to your brain. The brain considers the clues, compares them to your experience, and takes appropriate action.

All our senses have similar systems. Each has one or several types of receptor neurons that receive just one kind of environmental clue. In vision, light of certain wavelengths is converted into electric impulses in neurons in the eye. The other senses respond to vibrations in the air that enter the ear, chemicals in a liquid on the tongue or gas in the nose, or pressure on the skin. In each case, specialized receptor neurons change the environmental clue into a signal that travels to the brain. The brain sorts the signals into our perceptions of vision, sound, taste, smell, and touch.

Sensory Information

All sensory systems collect four types of information from environmental clues, or stimuli. One type of information is sensation. For humans these are vision, touch, taste, hearing, and smell. Each kind of sensation has several parts, such as color and movement in vision.

Another type of information is the amount of sensation. If there is not a large enough stimulus, the system does not detect anything. The amount of sensation that you can sense changes with different conditions.

Another type of information is how long the perception of the sensation lasts. If the stimulus lasts a long time, the amount of sensation decreases. For example, when you first get into a hot bath, the temperature might feel too hot, but this sensation fades quickly.

The last type of information is where the stimulus takes place. This affects the ability to distinguish two closely spaced stimuli. To go back to the ant example, if you have two ants on your arm very close together, can you tell if it is one or two ants? This depends on the number and density of the receptors. The more densely packed the receptors, the closer two ants can be and still be detected by two separate receptors.

Different areas of the brain process messages from different sensory systems. Where the brain receives the message determines whether we see or hear or smell something in response.

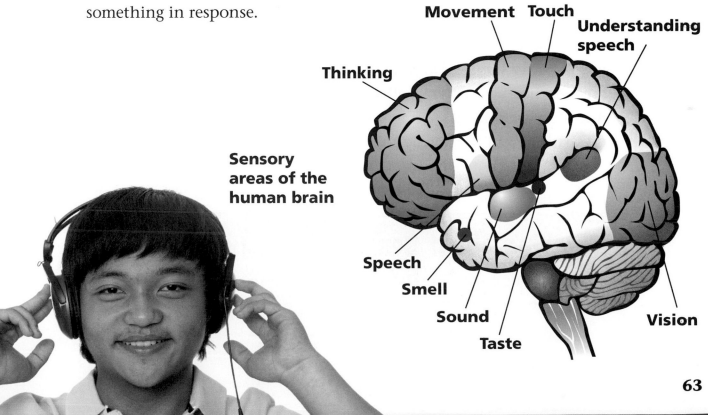

Sensory areas of the human brain

Movement Touch Understanding speech Thinking Speech Smell Sound Taste Vision

63

Animals respond to senses differently than humans.

Other animals have different ranges of sensory sensitivity than humans. Eagles see with greater acuity, bees see ultraviolet light, and rattlesnakes detect infrared, or heat. Dogs hear higher sounds and whales lower sounds. Dogs can smell thousands of times better, and great horned owls don't smell at all. And some animals, particularly migratory waterfowl, seem to sense magnetism, although the exact mechanism and the organs are unknown.

Thinking about Sensory Systems

1. How is our sense of vision like our sense of smell? How are they different?

Animal Communication

What is communication? It is passing information from one place to another. When you tell a friend you are thirsty, you are communicating a need for a drink.

Your thirst receptors tell your brain that you are thirsty. Your brain sends an action message to make a sentence. It also sends a message to your mouth to form the words "I am thirsty." If the spoken message enters your friend's ears, her sound receptors produce a signal. A sensory neuron takes the signal to her auditory center. In her brain's speech center, the sounds can be interpreted as a request for water.

A lot has to happen in this simple communication.

Sound

Many animals communicate using sound. Communication to others helps animals survive and reproduce. Wolves and coyotes howl to communicate with each other. Humpback whales make a loud whining sound, called singing. Their songs travel for many kilometers through the ocean. Their songs keep social groups in contact. The distinctive buzzing sound made by a rattlesnake warns potential predators to stay away. Prairie dogs make high-pitched chirps to communicate danger to other prairie dogs.

Other animals that use sound to communicate include crickets, frogs, lions, elephants, alligators, porpoises, owls, red-winged blackbirds—even shrimp. The pistol shrimp can produce one of the loudest sounds in the world.

A wolf and a humpback whale use sound to communicate.

A monarch butterfly and a peacock use the sense of vision to communicate in their environment.

Vision

Animals also use the sense of vision to communicate information to others in their environment. Vision can be used in a couple of ways. An animal might display itself to attract a mate or frighten a predator. The peacock's display is a magnificent fan of tail feathers. Its display is aimed at attracting a mate. The monarch butterfly's bright orange wings communicate to potential predators that it is toxic. Any predator that once tried to eat a monarch remembers the bright colors and avoids attacking a monarch again.

Other animals that use visual signals for communication include fireflies, hummingbirds, skunks, and birds of paradise.

Smell

Smell is a powerful communication tool in much of the animal world. Chemicals called pheromones produce odors that communicate important information. Some moths rely on pheromones to locate a mate. A mature female moth emits a small amount of the pheromone into the night air. When a male moth detects a single molecule of the pheromone, he starts to fly toward the odor. By following the scent, he might find a mate.

Ants use pheromones to mark their trails. Their pheromones can communicate the direction to a food source, the direction home, the presence of an enemy, or other valuable information.

All members of the cat family and dog family use scent markers. Dogs use urine to mark their territory. Chemicals in the urine identify the animal and tell its gender, age, and other information. The main message to outsiders is "Private property! Keep out!" Cats have scent glands at the corners of their mouth. They rub their face against structures that mark the boundary of their territory. They also mark objects within their territory. Does your cat rub its face on a chair leg or your pant leg? It is making sure that other cats know that it has claimed you as personal property.

Ants use pheromones to mark their trails.

A cat rubs its face against a fence to mark its territory.

A skunk uses odor as a weapon.

Odor can also be a weapon. The skunk is famous for its offensive odor. It can fend off most potential predators.

Some flowers have a peculiar odor. The cadaver flower smells like rotting, dead flesh. The odor attracts flies, because it seems like an excellent place to lay eggs. When the eggs hatch, the larvae should have plenty of rotting flesh to eat. But the smell is a false signal used to attract the fly so it will pollinate the flower. Many other plants also communicate with animals.

Humans' sense of smell is not as strong as that of many other animals. Yet it still helps us understand our surroundings. Next time you're walking down your street, pay attention to the odors. What do they tell you about the time of year, where you are, or what you should be cautious of?

Monarch Migration

The monarch butterfly is a migratory animal. It travels from the northern United States and southern Canada to Mexico, and then comes back. But no one butterfly makes the whole round trip. The migration system involves several generations of butterflies.

Let's start with a female monarch butterfly in Ohio. It's August. She hatched farther south, maybe on a farm in Kentucky. After she mates, she looks for a place to lay her eggs. She lays her eggs on milkweed because milkweed is the only food source for monarch larvae.

Soon after laying her eggs, the female butterfly dies. Her eggs hatch in a few weeks. The larvae eat milkweed leaves for several weeks, until they are about the size of your index finger. Then the monarch larvae pupate. They spend a few weeks inside the protective chrysalis. After the change from larvae to adult, each chrysalis splits open. The new adult butterflies climb out and pump up their new wings.

By this time, it is fall. The young monarchs have an adventure ahead. They must fly from Ohio to Mexico. Fall monarch butterflies instinctively start to migrate south all by themselves. The young monarchs head for a place they have never seen before. They take off without a map or a leader. They fly and fly, guided by instinct. Their need to migrate south is an **inherited trait** shared by all the fall monarchs.

Migrating monarchs from all over the eastern half of the United States end up in a small pine-oak forest in the mountains in central Mexico. Here the butterflies settle down with millions of other monarchs from the north. It is a safe place to spend the winter. The millions of monarchs crowd together for protection from predators and weather. They are inactive throughout the winter.

In spring, the days get longer and warmer. The butterflies become active, drink water, and start flying north. But the butterflies that migrated from Ohio do not fly back to Ohio. They fly as far as Texas, Louisiana, or Mississippi. They look for fields of milkweed and lay their eggs. The monarchs that made the long migration to Mexico die. Their offspring, generation 1, continue the northward migration.

When the weather gets cold, monarchs migrate from the eastern United States to central Mexico. They cluster in the pine-oak forest.

These first generation (Gen. 1) spring monarchs grow to be adults and fly a short distance north. They find milkweed, mate, lay eggs, and die. Their offspring, the second generation (Gen. 2), continue the northward migration. They find milkweed, mate, lay eggs, and die. Their offspring, generation 3, continue the journey.

The third generation (Gen. 3) monarchs fly north. They reach the limit of the monarch's range. By now, it is well into the summer. Some adult monarchs make it to Ohio, completing the Ohio monarch cycle. Others end up in Wisconsin, Michigan, or Maine. Some continue north into Canada. These third generation migrants find milkweed, mate, lay eggs, and die. By the time the fourth generation (Gen. 4) hatches, it is fall.

The adults in generation 4 live much longer than their parents or grandparents. They live for 6 to 7 months. They start to migrate south, responding to the shorter, cooler days of fall.

Look at the map of the monarch migration. Can you trace the northern migration, generation by generation?

The monarch butterfly migration system

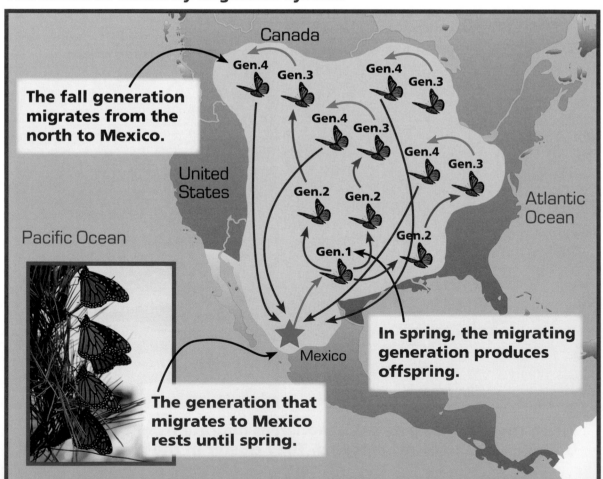

Monarch observers have recently seen many fewer monarch butterflies arriving in Mexico. The reasons for the decrease are not fully understood. One reason is a decrease in milkweed plants along the monarch's migratory path. Milkweed used to grow as a weed on farmland. Now farmers have better ways to prevent milkweed from growing. So monarchs have a harder time finding safe locations to lay eggs. The monarch larva depends on the milkweed plant for food. Any change in milkweed growth affects the survival of monarch populations.

How can you help the monarch populations? You can help teach people in your community about the migration system of monarchs and their need for milkweed. You can find out what native milkweed plants grow in your state and plant milkweed seeds. You can work with others to make sure milkweed plants thrive.

Thinking about Monarchs

1. Think about the monarch migration system. What are the parts?

2. What natural causes might affect the growth of milkweed plants? How might humans affect milkweed growth?

3. Predict the effect of logging in the pine-oak forest of central Mexico where monarchs spend the winter.

4. What are communities doing to protect monarchs?

North Atlantic Ocean Ecosystem

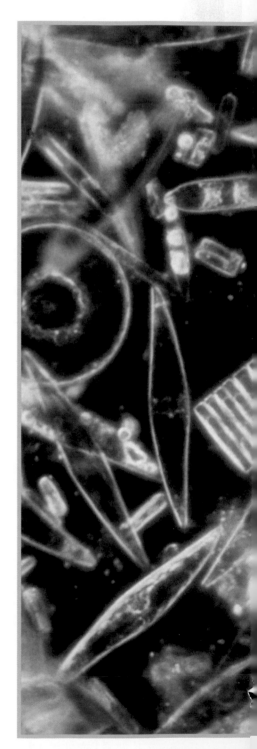

The largest ecosystems on Earth are the ocean ecosystems. Fish, crustaceans, mollusks, birds, and mammals of all sizes live and interact there. Like all ecosystems, ocean ecosystems have large populations of producers and consumers. Primary consumers feed on producers. Secondary consumers feed on primary consumers, and so on.

Where are the producers? You can see some of them on the ocean shores. Plant-like organisms called kelp and other seaweeds live in shallow water and near the ocean's surface. Seaweed cells contain chlorophyll, so they can make food from carbon dioxide (CO_2) during photosynthesis. Some shallow areas have sea grass, which also turns carbon dioxide into sugar during photosynthesis. Kelp and sea grass are two producers. But you probably have never seen the most important producers in the ocean. The base of the ocean food pyramid is microscopic organisms called phytoplankton.

There are thousands of species of phytoplankton. They are photosynthetic. They all need water, carbon dioxide, sunlight, and minerals. Water is all around organisms living in the ocean. Carbon dioxide from the atmosphere gets dissolved in the seawater. Sunlight can shine into the top few meters of the ocean (the photic zone).

The research vessel, Knorr, went on a 3-week expedition.

Phytoplankton Blooms

In spring, the number of hours of daylight increases. Conditions are right for large increases in the number of microscopic producers. These population growth spurts are called blooms. A bloom occurs when many phytoplankton reproduce and live and fewer die.

Scientists are studying the phytoplankton bloom in the North Atlantic Ocean. This bloom is important for two reasons. First, it produces the food that supports the North Atlantic ecosystem. This ecosystem produces fish that help feed people all over the world. Second, the phytoplankton remove tons of carbon dioxide from the atmosphere.

Researchers wanted to know what **variables** drive the annual phytoplankton bloom. They observed that the bloom starts weeks before the days start to get longer. But they didn't know why. The research team put robotic instruments into the North Atlantic early in the spring. The instruments collected data. The data showed that rough seas push cold northern water under warmer nutrient-rich water. As the cold and warm layers interact, eddies (whirlpools) form. The eddies bring phytoplankton up into the photic zone. These conditions encourage rapid reproduction of the phytoplankton. The eddies get the bloom started several weeks before the longer days of spring.

The North Atlantic bloom generates a huge amount of food for a diverse food web. How does the food web work? Zooplankton feed on the phytoplankton producers. Zooplankton are tiny (microscopic) animals. These include the larvae of crustaceans (such as crabs and shrimp), larvae of fish, larvae of jellyfish, marine copepods, and a host of other tiny animals. Zooplankton are eaten by larger animals, including fish, jellyfish, krill, seabirds, and whales. Just as terrestrial ecosystems depend on green producers such as grass and trees, ocean ecosystems depend on green phytoplankton.

A storm in the North Atlantic stirs up a large amount of food for seabirds.

A tiny marine copepod

The Carbon Cycle

Carbon is essential for plants and animals to survive. Terrestrial plants get carbon from carbon dioxide gas during photosynthesis. Phytoplankton get dissolved carbon dioxide from seawater. The carbon stays in organisms until it is digested for energy. Digestion breaks down the food and releases carbon dioxide into the environment. The carbon dioxide circulates in the atmosphere until an autotroph uses it or it dissolves into a body of water. The movement of carbon dioxide from the environment to organisms and back into the environment is the carbon cycle.

The carbon cycle

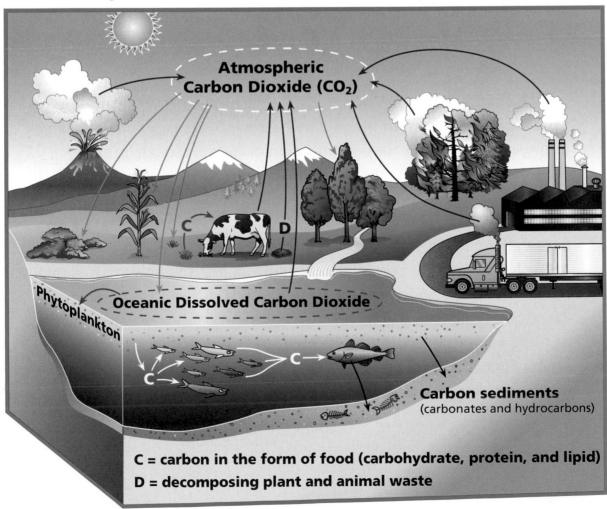

C = carbon in the form of food (carbohydrate, protein, and lipid)
D = decomposing plant and animal waste

Testing samples from the North Atlantic bloom

In the atmosphere, carbon dioxide is a greenhouse gas. Greenhouse gases absorb energy from the Sun and prevent the energy from escaping into space. The greater the concentration of carbon dioxide in the atmosphere, the warmer the atmosphere gets. Scientists are concerned that an increase of carbon dioxide in the atmosphere is raising the temperature of Earth. Global warming might stress ecosystems and cause the sea level to rise all around the planet.

The North Atlantic bloom scientists are looking into how phytoplankton blooms affect the concentration of carbon dioxide in the atmosphere. Researchers calculate that 33 percent of the carbon dioxide produced by burning fossil fuels is captured by phytoplankton. The North Atlantic phytoplankton bloom captures 20 percent of that. Some of the carbon drifts down into the deep ocean, so it stays out of the atmosphere.

Oceanic Research Instrumentation

The North Atlantic bloom research was a collaboration of many science specialists and engineers. To understand what was going on in the ecosystem, they needed to decide what data to collect and how to collect the data. Because the sea is so stormy and rough in the early spring, they could not simply lower their instruments into the water over the side of a ship. They needed to release instruments to float unattended for a period of time.

Two robotic instruments were designed to obtain data. One was a buoy that floats freely in the ocean. It collected data on current speed and direction, and temperature at and near the surface. The buoy stays at a planned depth not far beneath the surface. From anywhere in the ocean, communication antennae transfer data to a satellite that researchers on land can check any time. The buoy can make observations for weeks at a time before being retrieved.

Scientists prepare to lower the robotic buoy into the North Atlantic.

Knorr recovers the buoy.

Scientists collect the glider from the North Atlantic

This scientist pulls the water sampler back on the vessel to study.

Engineers prepare the seaglider for launch.

The second instrument was a robotic probe that swims. It is like a tiny submarine that carries instruments, not people. This robot can dive down to 1,000 meters (m) and then glide back to the surface. On each glide, instruments can monitor plankton, nutrients, temperature, and currents. The glider sends data to researchers on land via satellite.

Researchers on ships used other instruments to make observations. Special devices collected samples of seawater at different depths. These samples showed how many phytoplankton and zooplankton were at different depths.

In this research, data on two variables, time and place, are necessary. These data tell when the bloom starts, where it occurs, and when and where it ends. When researchers observe many phytoplankton dying, they know that the dead phytoplankton drift to the bottom of the sea. When this happens, the carbon in the phytoplankton can no longer cycle back into the atmosphere.

This kind of scientific discovery requires close collaboration. Scientists design the experiments, and engineers invent and build the amazing instruments that make observations and deliver data.

Science Safety Rules

① Listen carefully to your teacher's instructions. Follow all directions. Ask questions if you don't know what to do.

② Tell your teacher if you have any allergies.

③ Never put any materials in your mouth. Do not taste anything unless your teacher tells you to do so.

④ Never smell any unknown material. If your teacher tells you to smell something, wave your hand over the material to bring the smell toward your nose.

⑤ Do not touch your face, mouth, ears, eyes, or nose while working with chemicals, plants, or animals.

⑥ Always protect your eyes. Wear safety goggles when necessary. Tell your teacher if you wear contact lenses.

⑦ Always wash your hands with soap and warm water after handling chemicals, plants, or animals.

⑧ Never mix any chemicals unless your teacher tells you to do so.

⑨ Report all spills, accidents, and injuries to your teacher.

⑩ Treat animals with respect, caution, and consideration.

⑪ Clean up your work space after each investigation.

⑫ Act responsibly during all science activities.

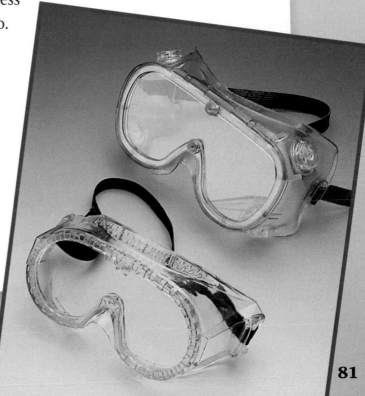

Glossary

aquatic referring to water

artery a blood vessel that carries blood from the heart to the body

atmosphere the layer of gases surrounding Earth

bacteria microorganisms that act as decomposers

biosphere a system of interacting living organisms on Earth

blade the flat part of a leaf

bolus a wad of food

brain part of the central nervous system protected by the skull

capillary the smallest blood vessel. Gases, nutrients, and wastes are exchanged between capillaries and cells.

carbohydrate a nutrient, such as sugar and starch, that provides energy

carbon dioxide (CO₂) a waste gas produced during cellular respiration. Plants use carbon dioxide during photosynthesis to make food.

cell the basic unit of life

central nervous system where sensory impulses pass through the brain and spinal cord

chlorophyll a molecule that absorbs red and blue light and reflects green light

circulate to move in a circle

circulatory system the system of blood vessels and organs that transports blood to all the cells in the body

classification the process by which scientists identify and organize objects and organisms, such as plants

classify to identify and organize according to similar properties or other criteria

colon the large intestine where solid waste is compacted in preparation for elimination

compete to rely on or need the same resource as another organism

consumer an organism that cannot make its own food. Consumers eat other organisms.

cytoplasm the liquid that fills living cells

decomposer an organism that breaks down plant and animal material into simple chemicals

detritivore an organism that feeds on broken down materials from dead organisms

detritus organic litter made up of dead organisms, their parts, and waste

digestion the process of breaking down food into nutrients that can be used by cells

digestive system the organs and structures that digest food. The digestive system includes the teeth, mouth, esophagus, stomach, small intestine, large intestine, and colon.

ecosystem a community of organisms interacting with each other and with the nonliving environment

energy what cells need to do work

esophagus the tube connecting the mouth and the stomach

food chain a description of the feeding relationships between organisms in an environment

food web the feeding relationships among all the organisms in an ecosystem. Arrows show the flow of matter and energy from one organism to another.

fungus (plural **fungi**) an organism that lacks chlorophyll and gets nutrients from dead or living organisms

geosphere Earth's core, mantle, and crust

glucose a sugar found in food; the sugar broken down in cells to release energy

heart a muscular organ that pumps blood

hydrosphere the interacting water on, under, and above Earth's surface

inherited trait a characteristic that is passed down from generation to generation

interact to act upon one another

large intestine the part of the digestive system between the small intestine and the rectum where water is removed from the solid waste

left atrium the upper chamber on the left side of the heart

left ventricle the lower chamber on the left side of the heart

lung the organ in animals where gases, such as oxygen and carbon dioxide, pass between the atmosphere and the blood

margin the edge of a leaf

membrane the outside of a living cell

mineral a nutrient that xylem transports to the cells in a vascular plant

motor neuron the cells that send information to the muscles

mouth a body opening where an animal takes in food

multicellular organism an organism composed of many cells

neuron a communication cell found in the brain and nervous system

nonliving referring to something that has never been alive or to things that were once alive and are no longer alive

nutrient a chemical in food that helps keep an organism alive and active

oxygen a waste gas produced by plants during photosynthesis. Oxygen is used by all plants and animals during cellular respiration.

palmate describing a leaf in which several veins start at one point near the base. The veins look like the fingers of a hand.

parallel describing a leaf in which the veins are straight lines all running in the same direction

phloem the long cells through which nutrients, such as sugars, are distributed in a plant

photosynthesis a process used by plants and algae to make sugar (food) out of light, carbon dioxide, and water

phytoplankton microscopic plantlike organisms in aquatic environments that produce their own food

pinnate describing a leaf that has one main vein with smaller veins branching off sideways from it

producer an organism, such as a plant or algae, that makes its own food

receptor cells that send messages to the brain when they receive stimuli

respiratory system the system of lungs and connecting tubes that transports oxygen to the red blood cells and gets rid of carbon dioxide

respond to react or to answer

right atrium the upper chamber on the right side of the heart

right ventricle the lower chamber on the right side of the heart

saliva the liquid produced in the mouth that aids digestion

sap a sugar-rich liquid transported by phloem

sensory neuron a nerve cell that sends information from sense organs to the brain

small intestine the part of the digestive system between the stomach and large intestine, where nutrients are absorbed from digested food

specialized structure a structure used primarily for one purpose

stimulus something that causes an action or response

stomach the organ where food is reduced to mush by acid and muscle activity

sugar the nutrient that cells use for energy

system a collection of interacting parts

teeth hard structures in the mouth used for cutting, biting, and chewing food

terrestrial referring to land

transpiration the process in which water is removed from the cells and passes into the environment

transport to move or carry

variable anything you can change in an experiment that might affect the outcome

vascular bundle the group of xylem tubes and phloem tubes in a vascular plant

vascular plant a plant with an internal system of tubes for transporting nutrients to its roots, stems, and leaves

vein a blood vessel that carries blood from the body to the heart

xylem the hollow cells of a plant that transport water and minerals to plant cells

zooplankton microscopic animals in aquatic environments

Index